Pet Detectives

Search for the Facts

Fish

Lindsey Lowe

BROWN BEAR BOOKS

Published by Brown Bear Books Ltd
4877 N. Circulo Bujia
Tucson, AZ 85718
USA

and

G14, Regent Studios
1 Thane Villas
London N7 7PH
UK

ISBN 978-1-78121-827-3

Library of Congress Cataloging-in-Publication Data available on request

Design Manager: Keith Davis
Children's Publisher: Anne O'Daly
Picture Manager: Sophie Mortimer

Picture Credits
Cover: Shutterstock: Sommai.
Interior: iStock: Iurii Bukhta 23, EuToch 1, 10, 22tl, CO Leong 21t, 22br, mtreasure 12, 22br, Andrey Nikitin 7t, Mirko Rosenau 3, 12–13, 14–15, 17, 18–19, slowmotiongli 16, Dmitri Sartcev 9, supermario 7cr, Vlad91 20t, 20–21; Shutterstock: ArtEventET 6–7, cynoclub 4-5, Tio Gian 14, Mikhail Grigorev 8–9, Songsak Pandet 4bl, Arunee Rodloy 21cr, Dmitri Startcev 5t, topimages 18, 22bl, Vangert 10–11.

All artwork and other photography Brown Bear Books

Brown Bear Books has made every attempt to contact the copyright holder.
If you have any information about omissions, please contact: licensing@brownbearbooks.co.uk

Manufactured in the United States of America
CPSIA compliance information: Batch#AG/5650

Websites
The website addresses in this book were valid at the time of going to press. However, it is possible that contents or addresses may change following publication of this book. No responsibility for any such changes can be accepted by the author or the publisher. Readers should be supervised when they access the Internet.

Contents

Which Fish? 4

What You Will Need 6

Fish Fact Files

White Cloud Mountain Minnow 8

Goldfish 10

Neon Tetra 12

Molly 14

Guppy 16

Zebrafish 18

In the Wild 20

Quiz 22

Useful Words 23

Find out More 24

Index 24

Which Fish?

Fish make great pets.
There are many types to choose from.
Which fish is right for you?

The Right Fish for You?

- Do you want cold freshwater or tropical fish?
- Do you have space for a fish tank?
- Can your family afford the tank?
- Do you have time to clean the tank regularly?
- Can your family afford to buy all the things you need for the tank?

Cold Freshwater or Tropical?

Goldfish are coldwater fish. They live in fresh water at room temperature. Tetras are tropical fish. They are from hot parts of the world. Their water needs to be kept warm.

Before Buying

Check out the fish in this book. Some need quite large tanks. Do not buy fish that will grow too big for your tank. Some fish need heaters and lights. Get the tank ready before you buy your pet.

Getting Along

Different types of fish can live together. Check which fish get along. Make sure the fish you choose can share their tank with others.

What You Will Need

Fish need plenty of room to swim.
You will need a large tank.
A filter keeps the water clean.

What you Need

- Large tank
- Heater and thermostat for tropical fish
- Water filter
- Water plants and fish tank decorations
- Fish tank gravel
- Gravel cleaner

The Tank

The tank should be as big as possible. Start with a 20-gallon (75-l) tank. Water from the faucet contains chemicals. Filter the water before using it.

Plants and Decorations

Put water plants and objects in the tank. These give fish hiding places. Buy plants and decorations made for fish tanks.

Heating

Tropical fish need warm water. Use a tank heater with a thermostat to heat the water. It should be between 75°F and 82°F (24°C and 28°C).

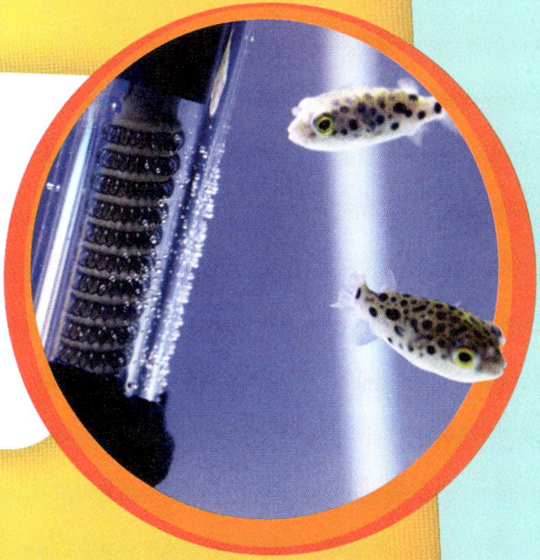

Water Filter

Fish cannot swim away from poop in the tank! Fit a special filter to keep the water clean. Change the filter regularly.

White Cloud Mountain Minnow

These fish are peaceful and social.
They can be kept with other fish.
Males are more colorful than females.

FACT FILE

Length:
1.5 inches (4 cm)

Food:
fish flakes, freeze-dried flakes of brine shrimp, and bloodworms

Color:
silver and bronze, shiny blue stripe, males have red markings

Lifespan:
4 to 5 years

WHERE IS IT FROM?

WOW!

White Cloud Mountain Minnows can live in water as hot as 72°F (22°C)!

These fish like to live in groups. They need a lot of space to swim around in.

Goldfish

Goldfish are easy to look after. They are happy on their own. They can recognize faces!

FACT FILE

Length:
2 to 4 inches (5 to 10 cm)

Food:
Goldfish pellets and flakes, chopped, shelled peas, lettuce, kale, spinach

Colors:
gold, orange, red, white, black, or a mix of white, orange and red

Lifespan:
7 to 10 years

Goldfish often pick up bits of gravel. They suck off tiny creatures. Then they spit the gravel out.

WOW!

Most fish have no eyelids. They sleep with their eyes open!

WHERE IS IT FROM?

Neon Tetra

Neon Tetras live in warm water.
They are small and active.
They are peaceful and shy.

FACT FILE

Length:
1.5 inches (4 cm)

Food:
fish flakes and pellets, brine shrimp, fruit flies, shredded cabbage

Color:
neon blue, bright red stripe, silvery-white

Lifespan:
3 to 5 years

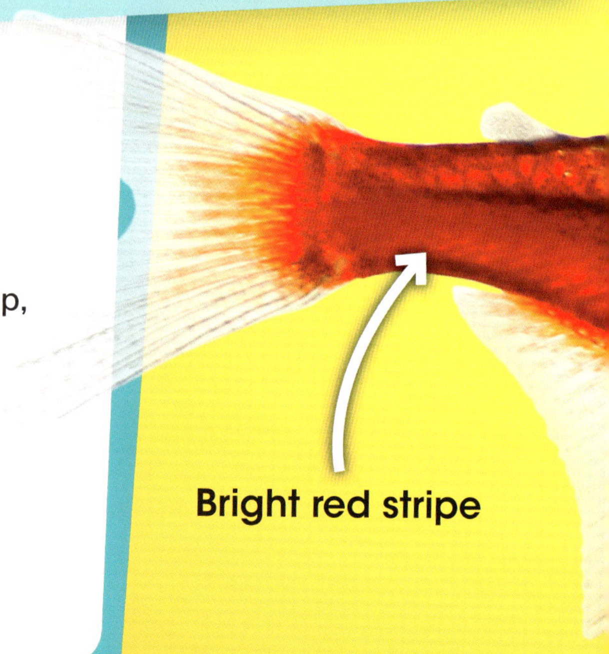

Neon Tetras are "schooling" fish. That means they like to live in large groups. They need large tanks.

WOW!

The bright colors fade at night. They brighten again when the Neon Tetra wakes up!

WHERE IS IT FROM?

Molly

Mollies like to hide behind plants. They will explore the tank together. They are fun to watch!

Fan-shaped tail with points top and bottom

FACT FILE

Length:
3 to 5 inches (8 to 12 cm)

Food:
fish flakes and pellets, brine shrimp, algae, cooked lettuce, and spinach

Colors:
lots, including gold, black, white, gray, white, silver

Lifespan:
2 to 5 years

Molly fish come in lots of colors. Some are gold-and-black. Others are black-and-white or black.

WOW!

Mollies should not be overfed. They never know when to stop eating!

WHERE IS IT FROM?

Guppy

Guppy fish are active swimmers. They need plenty of space. They love to nibble on plants!

FACT FILE

Length:
1 to 2 inches (2.5 to 5 cm)

Food:
fish flakes, pellets, plants, brine shrimp, bloodworms, mosquito larvae

Colors:
lots, including silver, red, yellow, blue, purple, orange

Lifespan:
2 to 3 years

WOW!

Guppies give birth to live young. Females can give birth to around 50 baby fish every month!

Guppies have amazing tail fins. They are long and flowing. They come in many colors.

WHERE IS IT FROM?

Zebrafish

Zebrafish are good pets.
They are small and cute.
They have stripes like a zebra!

FACT FILE

Length:
1 to 2 inches (2.5 to 5 cm)

Food:
fish flakes and pellets, brine shrimp, bloodworms,

Colors:
silver or gold bodies, blue, purple, black, green, pink stripes

Lifespan:
3 to 5 years

There are several types of Zebrafish. Some have long fins. Others have short fins. Some are pink.

WOW!

In September 2014, some Zebrafish were sent into space. They were studied by space scientists!

WHERE IS IT FROM?

In the Wild

Wild fish are found all over the world. They live in fresh water or salt water. They come in different sizes and colors.

Bluefaced Angelfish

Bluefaced angelfish live in warm oceans. They live alone or in pairs. They like to hide. The young fish hide in dark caves.

Coral Reefs

Coral reefs are in warm, shallow water. They are home to lots of fish. The reef looks like rocks. But it is made of living creatures.

South America

Many aquarium fish come from South America. Red-Bellied Piranha swim in the Amazon River. They are sometimes kept in aquariums.

African Lake

These are Malawi Golden Cichlids. They are freshwater fish. In the wild, they only live in one place. That is Lake Malawi in Africa. The males fight in a tank. It's best just to keep one.

Quiz

Test your fish detective skills!
Can you answer these questions?
The answers are on page 24.

1 Do Goldfish and other fish have eyelids?

2 What happens to Neon Tetras at night?

3 Which type of fish was sent into space in 2014?

4 Where do these fish come from?

Useful Words

algae
Plantlike living thing, such as pond scum, seaweed, and moss.

aquarium
A glass tank for fish and other animals that live in water.

dorsal fin
The fin on the backs of fishes, whales, dolphins, and porpoises.

fins
The thin parts attached to the body of a fish. Fins help the fish to move through the water.

larvae
Young wingless forms of insects that hatch from eggs.

lifespan
The number of years something lives for.

organism
A living thing made up of one or more cells.

thermostat
A device that automatically controls temperature.

Find out More

Books

Caring for Fish, Tammy Gagne, Pebble Books, 2018

Is a Fish a Good Pet for Me?, Jason Brainard, PowerKids Press, 2020

Websites

www.coolkidfacts.com/fish-facts/

www.ducksters.com/animals/fish.php

www.petsathome.com/pet-talk/choosing-a-pet-fish-fish-type

Index

aquarium 5, 21, 23

filter 6, 7

fins 8, 13, 16, 18, 23

heater 5. 6. 7

plants 6, 7

schooling fish 12, 20

tank 4, 5, 6, 7, 12, 14

temperature 5, 7, 9

thermostat 6, 7, 23

wild fish 20, 21

what you will need 6. 7

Quiz Answers: 1. No. Most fish have no eyelids. **2.** Their bright colors fade. They come back when the Neon Tetra wakes up. **3.** Zebrafish. They were studied by space scientists. **4.** The Amazon River in South America.